MW01222391

GRASSHOPPERS

Published by Smart Apple Media
123 South Broad Street
Mankato, Minnesota 56001

Printed in the United States of America.

Photos: Whitney Cranshaw (pages 15–16, 19, 28, 30); Howard Ensign Evans (pages 22, 25); John Capinera (page 8); Dan L. Perlman (page 7); Entomological Society of America/Ries Memorial Slide Collection (pages 13–14, 20–21, 23, 26–27); PhotoDisc (cover, pages 2–3, 8, 10–12, 17–18, 24, 29)

Design &Production: EvansDay Design
Project management: Odyssey Books

Library of Congress Cataloging-in-Publication Data
Halfmann, Janet, 1944–
Grasshoppers / Janet Halfmann. – 1st ed.
p. cm. – (Bugs)
Includes bibliographical references and index.
Summary: Describes the habitat, life cycle, behavior, predators, and unique characteristics of grasshoppers.
ISBN 1-887068-34-1 (alk. paper)
1. Grasshoppers: Juvenile literature. [1. Grasshoppers.]
I. Title. II. Series: Bugs (Mankato, Minn.)
595.7'26—dc21 98-15342

First Edition 9 8 7 6 5 4 3 2 1

GRASSHOPPERS
Janet Halfmann

A farmer watched helplessly as millions of black insects covered his wheat field. By dark, the wheat was *eaten to the ground!* The year was 1848 in what is now Utah. The Mormon settlers needed the wheat to keep from starving. *Suddenly,*

SEAGULLS FROM NEARBY ISLANDS IN THE GREAT SALT LAKE *swooped* IN TO FEAST ON THE INSECTS, SAVING MANY OF THE WHEAT FIELDS. TODAY, THESE BLACK INSECTS ARE KNOWN AS MORMON CRICKETS OR WESTERN GRASSHOPPERS (ANABRUS SIMPLEX).

The Grasshopper's Family

Fortunately, the grasshopper can be our friend as well as our enemy. It belongs to the ORDER, or group, of insects called *Orthoptera*, which means "straight-winged." The name refers to the straight front wings of most members of the order. Some grasshoppers, like the Mormon cricket, have tiny wings or no wings at all. This order also includes the true cricket, cockroach, and praying mantis.

Most grasshoppers in North America are ¾ to 3 inches (19–76 mm) long. The

Topidacris grasshoppers, which live in South America, are 6 inches (15 cm) long and have a wingspan of 10 inches (25 cm). Sometimes people mistake them for birds!

Short-horns and Long-horns Most grasshoppers belong to one of two families. The largest and best-known family is the *Acrididae*, or SHORT-HORNED grasshoppers. They get their name from their short, thick antennae, or feelers. This family includes the locusts and has about 10,000 SPECIES, or kinds. The second family is the *Tettigoniidae*, or LONG-HORNED grasshoppers. They have long, slender antennae, usually longer than their bodies. This family includes grasshoppers called katydids and Mormon crickets. It has about 5,000 species.

Grasshoppers live on plants all over the world. More than 600 species are in North America.

Old as the Dinosaurs

Grasshoppers are one of the oldest insects. Their ancestors lived on earth at the time of the dinosaurs 300 million years ago. The types of grasshoppers we see today have been around for 185 million years.

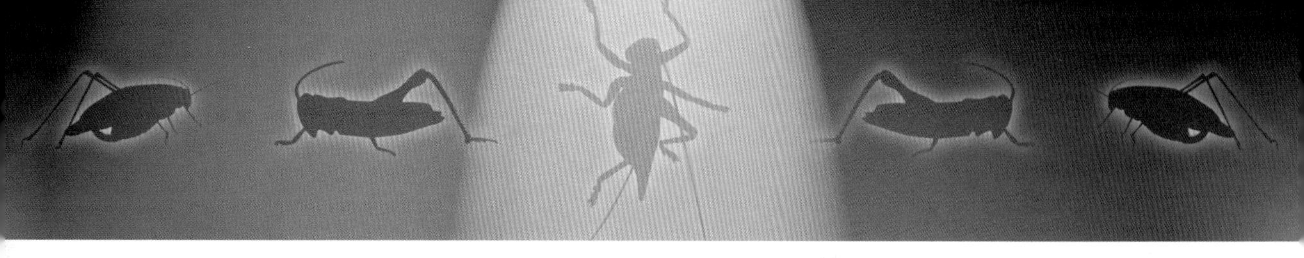

Habitat

Grasshoppers live all over the world where there is warm weather and plants for food. They are one of the most abundant insects. More than 600 species live in North America. A common short-horn is the little red-legged grasshopper *(Melanoplus femur-rubrum)*, which sports bright red hind legs. Also common is the large Carolina locust *(Dissosteira carolina)*, sometimes called the "dusty-road grasshopper" because it is so often seen alongside dusty roads.

One of the most familiar long-horned grasshoppers is the true katydid *(Pterophylla camellifolia)*. This large green grasshopper is famous for the male's loud

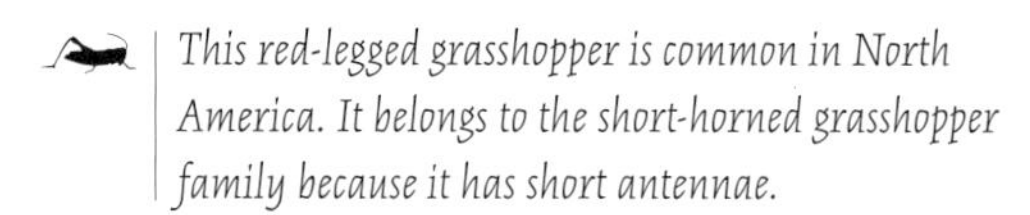

This red-legged grasshopper is common in North America. It belongs to the short-horned grasshopper family because it has short antennae.

This katydid is a long-horned grasshopper. Like many katydids, it looks a lot like a leaf.

song at dusk, "Katy-did, Katy-didn't." Other common long-horns are the meadow katydids named *Orchelimum*. They live in wet grassy areas and the tall grass of roadsides.

Dinosaur Grasshopper One short-horned grasshopper has the nickname "Dinosaur Grasshopper." It has a crest along its back similar to that of the fin-backed dinosaur Dimetrodon. Also known as the great

crested grasshopper *(Tropidolophus formosus)*, it lives in the southwestern United States.

Most short-horned grasshoppers live near the ground in meadows, fields, deserts, grasslands, and gardens. They like places with both plants and bare patches of earth where they can lay their eggs or bask in the sun. Many long-horned grasshoppers, such as katydids, live in leafy trees and shrubs.

Love Those Plants Both adult and baby grasshoppers live in the same places and eat a large variety of plants—everything from grasses and flowers to crops, vegetables, and

Both long-horned and short-horned grasshoppers have big appetites and eat a large variety of plants.

Basking in the Sun

Most grasshoppers are active during the day because they need to be warm to move, jump, and fly. You'll often see them basking on bare patches of earth, soaking up the warmth from the ground. But a few kinds of grasshoppers, especially long-horns like the katydids, are active at night.

fruit trees. Both have huge appetites. Grasshoppers have been known to wipe out whole fields of alfalfa, cotton, corn, or other grains! A few kinds of long-horned grasshoppers, like the great green grasshopper *(Tettigonia viridissima)*, also eat insects and the remains of animals.

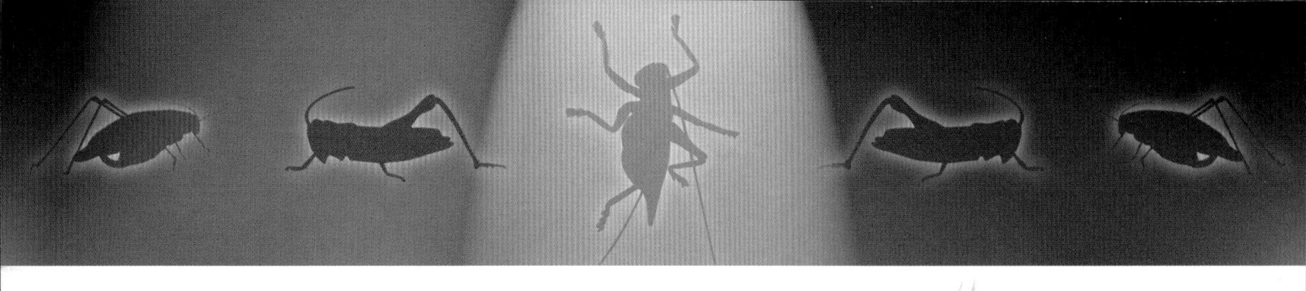

Body and Senses

The grasshopper likes to be clean. It washes its face like a cat with its front feet. To clean an antenna, it drags it under a foot. It polishes its big hind thighs by rubbing the bent knees of the middle legs against them. Like all insects, the grasshopper has a tough outer covering called an EXOSKELETON, which provides the body with support and protection. Also, like all insects, the grasshopper has

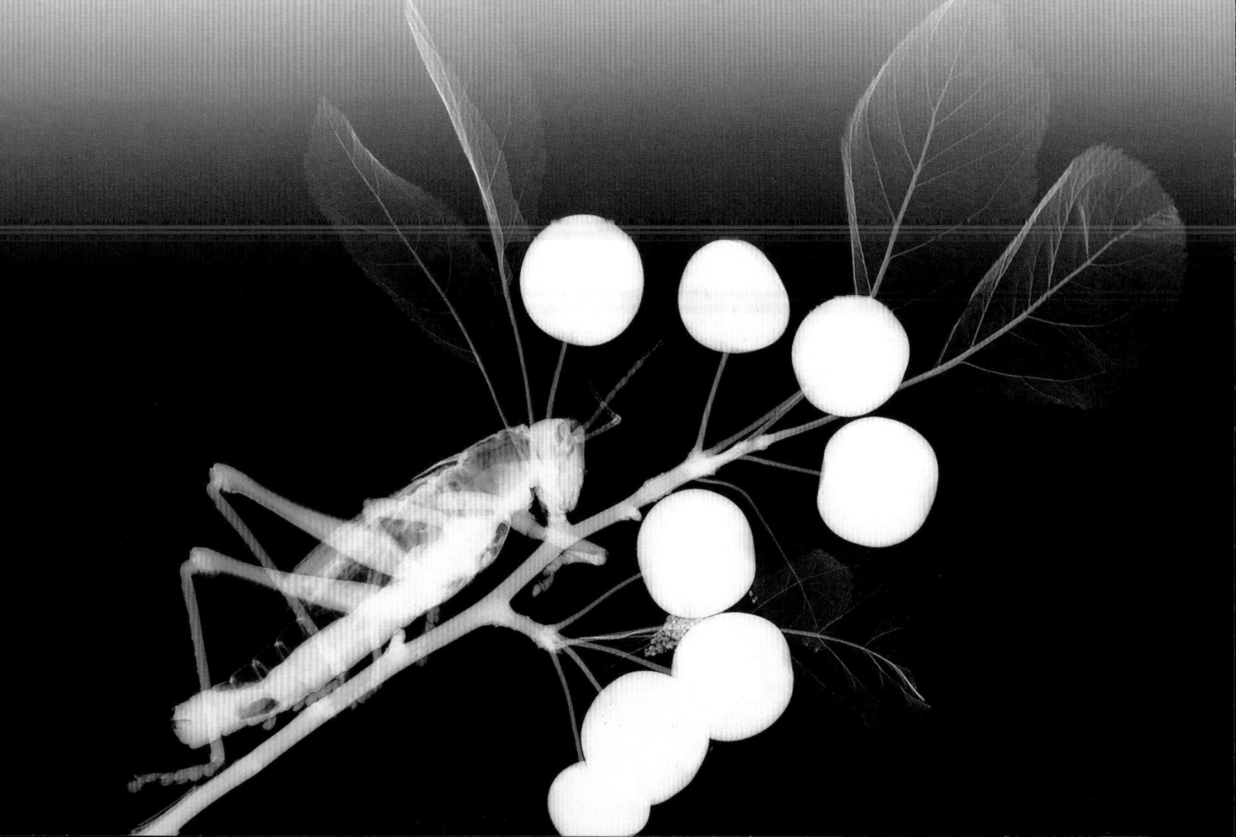

six legs and breathes air through tiny openings in the exoskeleton. Its body has three parts: head, thorax or chest, and abdomen.

Large Head with Big Mouth On the sides of the grasshopper's large head are two big eyes that can see in all directions at once. The grasshopper also has three tiny eyes in the center of its head that see only light. Most grasshoppers have keen vision.

You can't miss the grasshopper's large mouth. It has two big horny lips and sharp jaws with saw-like edges. Tiny feelers on the mouth taste the food. Two ANTENNAE, or feelers, stick out from the grasshopper's head. They are used for smell, touch, and sometimes hearing.

All grasshoppers have a tough outer covering called an exoskeleton and three body parts: head, thorax, and abdomen. This picture is an X ray.

Mighty Jumpers!

Imagine being able to leap over a tall building in a single bound. Superman could do it. So could a grasshopper if it were human sized. In fact, grasshoppers can jump 20 times the length of their bodies! If you were a 5-foot (1.5-m) tall grasshopper, that means you could jump 100 feet (30 m)! How do grasshoppers do it? It's all in their long hind legs.

Thorax The THORAX supports the legs and contains the muscles that power the wings. The grasshopper uses all six legs for walking, its main way for getting from place to place. It jumps and flies mostly to escape enemies. The front legs are also used to hold food. In long-horned grasshoppers, the TYMPANA, or hearing organs, are located on the lower front legs.

The grasshopper's long hind legs stick up high above its body. They are powered by big, strong muscles in the upper legs. In

fact, the grasshopper has the most powerful leg muscles in proportion to its size of any animal except the flea.

To jump, the grasshopper digs spines near its feet into the ground to get a good grip. Then it brings its upper and lower hind legs close together, thrusts them out straight, and shoots forward like a released spring.

Most grasshoppers have two pairs of wings. The front wings are narrow and tough, the rear ones wide and filmy. The rear pair are the main flying wings. They fold up like fans under the protective front wings when at rest.

Many kinds of grasshoppers are not good fliers. They use their wings like a glider to be able to jump farther. But some species, like the American bird grasshopper *(Schistocerca americana)*, can

Lubber grasshoppers like this one are some of the largest grasshoppers. They have short wings and cannot fly.

In long-horned grasshoppers like this one, the hearing organs are located on the lower front legs.

fly fast and far. True locusts also have powerful wings.

Long, Narrow Abdomen In short-horned grasshoppers, the TYMPANA (ears) are located on the abdomen just above the hind legs. The grasshopper's long, narrow ABDOMEN also contains the heart, digestive and breathing systems, and reproductive organs.

This Carolina locust lives along dusty roads, where she can blend in with the brown earth. Here she is laying her eggs.

The best way to find grasshoppers in the tall grass is to listen for their songs.

The Grasshopper's Songs

Imagine being in a field of tall grass. Your friends are in the field, too, but you can't see them because the grass is taller than all of you. Pretend you are a grasshopper and can't yell to your friends. How would you find one another?

You could put on a concert. That's what the grasshoppers do. They are famous for their buzzing or chirping songs. They rub together parts of their bodies to make sounds. This is called STRIDULATION. It's similar to drawing a bow across a violin.

Playing the Violin Most short-horned grasshoppers rub tiny pegs on the inside of the hind leg (the bow) over hard ridges on the front wing (the violin). Most long-horned grasshoppers, such as the katydids, rub a row of bumps on one front wing against a ridge on the other front wing.

Males do most of the singing. Some females sing, but their chirp is softer and is usually made in answer to the male. Some species, like the short-horned grasshoppers named *Parapodisma*, don't make sound at all.

Grasshoppers chirp to warn of danger or to tell others to stay away from their territory. One of the main reasons males chirp is so females can find them in the tall grass when it's time to mate.

Rock 'n' Roll or Beethoven? Listen carefully as you walk near a grassy field on a warm summer day. You may hear the grasshoppers chirping. Listen even more closely and you'll notice that not all the chirps are the same. Each species has its own unique songs, depending on the message.

For example, the male stripe-winged

Grasshoppers "sing" to one another by rubbing parts of their bodies together to make chirping sounds.

grasshopper *(Stenobothrus lineatus)* chirps a constant warbling note to attract a female, then, switches to a series of clicks to court her. The male *Omocestus viridulus,* or common green grasshopper, chirps a constant trill to attract a female, while the male common field grasshopper *(Chorthippus brunneus)* chirps only a single pulse every now and then. The different songs are important because only a male and a female of the same species are able to mate.

Love Song

Most species of grasshoppers mate in the late summer and early fall. When the female hears the male's song, she follows the sound until she finds him. She may chirp a soft answer. When the female comes near, the male strokes her with his antennae. In species that don't sing, the male and female find each other by sight and smell. The two join their abdomens and the male passes white bags filled with sperm along to the female. After about a week, the female is ready to lay her eggs.

Life of the Grasshopper

The grasshopper's life has three stages: egg, nymph, and adult. This process is called incomplete METAMORPHOSIS, or incomplete change. Most insects, like beetles, go through a four-part change that includes a pupal stage. This is called complete metamorphosis.

Eggs Wait Out the Winter Most short-horned grasshoppers lay their eggs in the

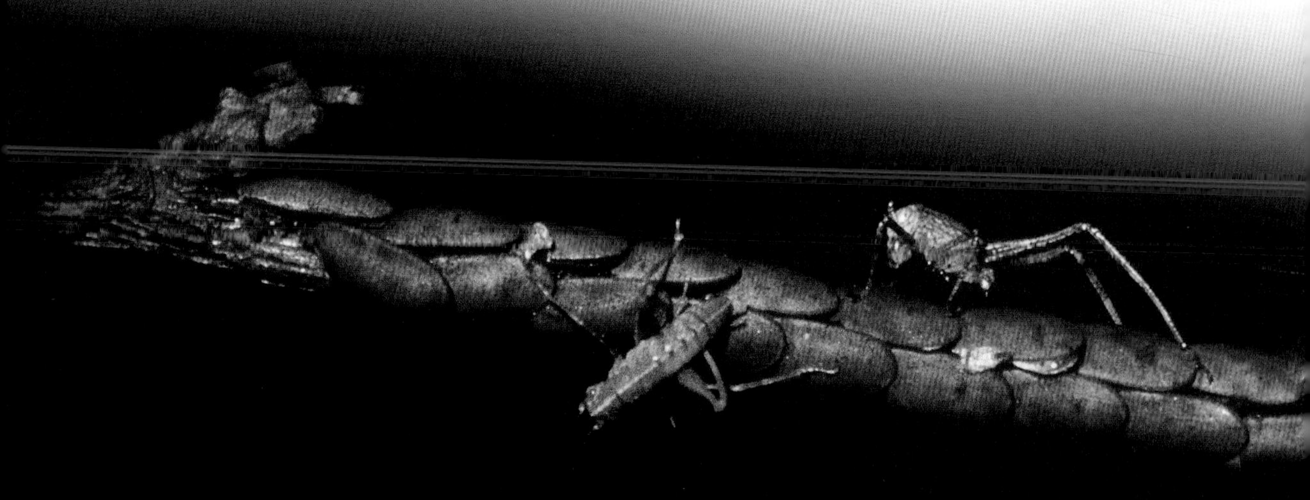

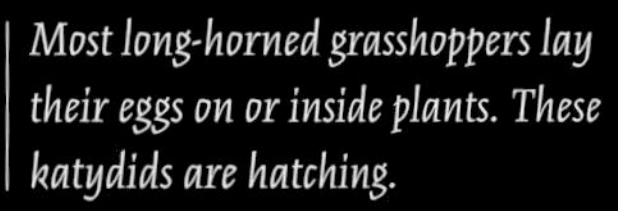

Most long-horned grasshoppers lay their eggs on or inside plants. These katydids are hatching.

soil. The female uses the tip of her abdomen as a drill to dig a hole as deep as she can. Then she drops a mass of eggs from the tip of her abdomen into the hole.

The number of eggs varies, depending on the species. For example, the painted grasshopper *(Dactylotum bicolor)* of the southwestern desert of the United States lays a mass of 100 eggs, while the red-legged grasshopper lays only about 20.

The female covers the eggs with sticky foam, which hardens around the eggs to form a pod that keeps them safe and dry. Then she fills the hole with dirt. She digs several more holes, and fills them with egg masses.

Some long-horned grasshoppers lay

This two-striped grasshopper, like most short-horns, pushes the tip of her abdomen into the soil to lay her eggs.

their eggs on or inside plants. Others lay them in the soil. They lay their eggs one at a time, and do not cover them with foam.

The new babies start to grow inside the eggs, but stop once winter comes. They will stay snug under the ground during the winter and hatch when the weather gets warm.

The Hungry Nymph In spring or summer, the baby grasshopper pushes out of the egg's shell. At first, it's wrapped in a skin that holds its legs and antennae against its body. Like a worm, it wriggles upward through the foam and dirt into the daylight.

Once out of the ground, the baby sheds its protective skin. Now it's called a NYMPH and can stand on its legs. It looks like its parents, except that it doesn't have wings. But it is already a good jumper.

The nymph is very hungry. It eats twice its weight every day! About once a week its skin gets too tight. Then it twists until its skin splits, and crawls out of it wearing a new larger skin. This process is called MOLTING. The young grasshopper sheds

Grasshopper nymphs, like this one, look like their parents, but they're smaller and don't have wings.

its skin six times. After the fourth molt, wing stubs appear that grow larger with each new skin.

Wings at Last When the nymph sheds its skin the last time, it is an adult. It unfolds its wrinkled wings and stretches them. In a few hours, they will harden and dry. Throughout the rest of the summer, the adult grasshopper's body continues to develop. Its muscles grow stronger so it can escape from its many enemies. But even if it is lucky enough to survive the summer, most probably it will die in the fall when it gets cold.

The last time a nymph sheds its skin, its wings are fully grown. This long-horned grasshopper has just become an adult.

Grasshopper Enemies

Grasshoppers face danger even before they emerge from the egg. Many eggs are eaten by insect larvae, such as blister beetles. A very wet spring can destroy many grasshopper eggs by causing the pods to mold.

Once out of the egg, jumping high, far, and fast is the grasshopper's main way to escape enemies. These PREDATORS include wasps, birds, lizards, frogs, foxes, skunks, cats, and mice. In fact, the grasshopper mouse is named after its favorite food.

Baby grasshoppers are good jumpers

except when they are molting—a dangerous time for them. Unfortunately for the grasshopper, jumping to safety doesn't always work. Sometimes a grasshopper leaps right into a spider's web!

Color Tricks The band-winged grasshoppers, such as the Carolina locust, have a special way of fooling their enemies. First, they startle them with the bright colors on their wings as they take off. When the enemy follows, the grasshopper suddenly drops to the dust. It folds its bright wings under its dull ones, becoming invisible, and the enemy shoots by.

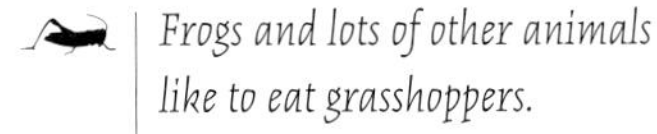
Frogs and lots of other animals like to eat grasshoppers.

Blending In

Many grasshoppers escape their enemies by blending in with their surroundings. This is called CAMOUFLAGE. Most grasshoppers are green or brown, making them difficult to see among the grasses and patches of earth where they live. Katydids often have wings shaped like leaves, complete with the veins.

Many grasshoppers have stripes to make them look even more like blades of grass. One of the best grass mimics is a grasshopper named *Cylindrotettix* from Brazil. It is shaped like a grass stem and has two different colors—green and straw—depending on the season.

Other grasshoppers use bright colors to warn enemies to stay away. In Africa, the common milkweed grasshopper *(Phymateus morbillosus)* shows off its crimson wings spotted with yellow. If that doesn't work, it squirts a stinky foam.

Grasshoppers can also kick with their powerful and often spiny hind legs. Some can bite. As a last resort, a grasshopper can shed a leg to get away.

Sometimes an insect's bright color is a warning for enemies to stay away. But in the case of this unusual pink katydid, its color is a chance happening or mutation. A pink katydid is more likely to be eaten because it doesn't blend in with its surroundings.

This grasshopper may be this flower's pal. Scientists have found that grasshopper spit makes plants grow faster.

Grasshoppers and Us

The grasshopper's huge appetite has caused problems with humans since ancient times. Just six or seven grasshoppers per square yard (.836 sq m) in a pasture can eat as much alfalfa hay as a cow. Grasshoppers have even been known to eat straw hats and handles of pitchforks left in fields when food is scarce!

Swarming Locusts The worst trouble has come from certain short-horned grasshoppers called locusts. From time to time,

they hatch in much larger numbers than usual. Then they grow longer wings and often take on bright colors. They band together and fly hundreds of miles to find food. A swarm may contain billions of grasshoppers. Wherever they land, they devour all the plant life for miles around.

Six of the most damaging species of locusts live in Africa. One swarm of the desert locust *(Schistocerca gregaria)* numbered 40 billion and measured 400 square miles (1,036 sq km) across! A swarm this size needs 89,000 tons (80,723 MT) of food per day, or enough to keep 400,000 people alive for a year.

Fortunately, only about 20 of the thousands of species of grasshoppers cause major damage. In North America, the two most serious pests are the migratory grasshopper *(Melanoplus sanguinipes)* and the red-legged grasshopper.

The migratory grasshopper shown here eats farm crops. It is one of North America's worst grasshopper pests.

In many countries, grasshoppers are considered a nutritious and tasty snack. They have a nutty flavor.

Yum-Yum, Roasted Grasshoppers! Not everyone lets the grasshopper do all the eating. In many countries, grasshoppers are considered an important source of protein. In fact, the grasshopper is the most widely eaten insect in the world.

Grasshoppers have a nutty taste and can be cooked in many ways. In Africa, they are roasted or ground into flour. In Mexico, mashed grasshoppers are mixed with beans and spread on tortillas. In Japan, grasshoppers are boiled in soy sauce. In

Grasshoppers can be pests when there are too many in one place. But their songs are a treat on a summer's day.

Sweden, some mint lollipops have roasted grasshoppers in the center!

Pal or Pest? Millions of swarming grasshoppers are definitely a problem. Sometimes the largest swarms have to be tracked down by airplanes and sprayed with poison. Yet one grasshopper nibbling on a plant may actually be its pal. Scientists have found that grasshopper spit makes plants grow faster.

Pal or pest, one thing is for certain. The warm days of late summer wouldn't be the same without a chorus of grasshoppers chirping all around us.

LEARNING RESOURCES

BOOKS

Biology of Grasshoppers, R.F. Chapman (Division of Neurobiology), John Wiley & Sons, 1990

Chirping Insects, Sylvia A. Johnson, Lerner Publications Company, 1986

Discovering Crickets and Grasshoppers, Keith Porter, The Bookwright Press, 1986

The Grasshopper, Yo Hasegawa, Raintree Publishers, 1986

Grasshoppers, Valerie K. Brown, Naturalists' Handbooks 2, Press Syndicate of the University of Cambridge, 1983

Grasshoppers, Jane Dallinger, Lerner Publications Company, 1981

Grasshoppers and Crickets, Dorothy Childs Hogner, Thomas Y. Crowell Company, 1960

Grasshoppers and Mantids of the World, Ken Preston-Mafham, Facts on File, 1990

Grasshoppers and Their Kin, Ross E. Hutchins (entomologist), Dodd, Mead & Company, 1972

Katydids, the Singing Insects, Barbara Ford, Julian Messner, 1976

Red Legs, Alice E. Goudey, Charles Scribner's Sons, 1966

CHAPTERS IN BOOKS

"Grasshoppers," *Ninety-nine Gnats, Nits, and Nibblers*, May R. Berenbaum (entomologist), University of Illinois Press, 1989

"Grasshoppers, Katydids & Crickets," *The Practical Entomologist*, Rick Imes, Simon & Schuster, 1992

WEB

"Field Guide to Common Western Grasshoppers," Robert E. Pfadt, Wyoming Agricultural Experiment Station Bulletin 912, September 1994

"Grasshoppers and Crickets (Orthoptera)," Gordon's Entomological Pages

"The Hills Are Alive with the Sounds of… Hungry Grasshoppers," Dr. Bug (entomologist)

"Insects in the Human Diet," Mad Science, October 1995

ENCYCLOPEDIAS AND REFERENCE BOOKS

Academic American Encyclopedia, 1992

The Bug Book, The 1981 Childcraft Annual, World Book–Childcraft International, Inc., pp. 84–93

Compton's Encyclopedia online

Encyclopedia Americana, 1995

Encyclopedia Britannica, 1993

The Encyclopedia of Insects, edited by Christopher O'Toole, Facts on File, 1986

Grolier online

Knowledge Adventure, 1997 online

National Audubon Society Field Guide to North American Insects & Spiders, 1995

World Book Encyclopedia, 1996

MAGAZINE ARTICLES

"Bug Juice: A Nutritious Plant Food?" *Science News*, June 17, 1995, p. 381

"Bugs for Breakfast," Peggy Thomas, *Cricket*, September 1996, pp. 11+

"Candy That Hops off the Shelves," *Maclean's*, October 23, 1995, p. 7

"Extra Cheese and Bugs to Go!" Deborah Churchman, *Ranger Rick*, January 1995, pp. 32+

"A Never-Ending Feast," *Scientific American*, November 1995, p. 24

"Plague and Pal," *Discover*, October 1995, p. 33

MUSEUMS

California Academy of Sciences
Golden Gate Park
San Francisco, CA

The Milwaukee Public Museum
Milwaukee, WI

Natural History Museum of Los Angeles County
Los Angeles, CA

Smithsonian Institution
Washington, DC

Students' Museum, Inc.
Knoxville, TN

INDEX